ε Epsilon

Focus: Fractions

Test Booklet

Math·U·See

1-888-854-MATH (6284)
www.MathUSee.com

Math·U·See

1-888-854-MATH (6284)
www.MathUSee.com
Copyright © 2009 by Steven P. Demme

Build the problem, then write the correct solution.

1. **Step 1** Select 12 blocks.

   Step 2 Divide into 2 equal parts.

   Step 3 Count 1 of them.

   —— of ____ is ____

2. **Step 1** Select 16 blocks.

   Step 2 Divide into 4 equal parts.

   Step 3 Count 3 of them.

   —— of ____ is ____

Read the problem, build it, and write the answer in the blank.

3. One-third of fifteen is _____.

4. Two-fifths of twenty is _____.

Solve.

5. $\dfrac{1}{8}$ of 16 = _____

6. $\dfrac{3}{4}$ of 8 = _____

7. $\dfrac{1}{2}$ of 10 = _____

8. $\dfrac{3}{6}$ of 36 = _____

9. $\dfrac{2}{7}$ of 49 = _____

10. $\dfrac{4}{5}$ of 40 = _____

Find the perimeter of each rectangle.

11.
5"

8"

P = _____

12.
12'

24'

P = _____

13.
32"

55"

P = _____

14. One-sixth of the people in the room had blue eyes. If there are 18 people in the room, how many have blue eyes?

15. Kimberly bought a book that cost $14.58 and a box of paints for $6.75. How much did she spend?

16. Kimberly had $25 before she went shopping (#15). How much money did she have after buying her book and paints?

17. Allison went on a two-week vacation. If it rained 2/7 of the days she was on vacation, how many days did it rain?

18. Raleigh built a yard for his dog. The yard was a rectangle that measured 12 feet by 15 feet. How many feet of fence did Raleigh need to buy?

2

Write in the correct numerator and denominator in both symbols and words.

1.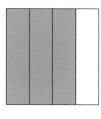

_____

_____

2.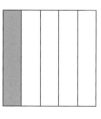

_____

_____

3.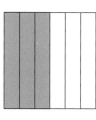

_____

_____

Build the fraction, then draw it in the square. Say it and write it with numerals or words.

4.

_____

four-fifths

5.

_____

one-third

6.

$\frac{1}{2}$

_____

Solve.

7. $\frac{2}{5}$ of 25 = _____

8. $\frac{1}{10}$ of 80 = _____

9. $\frac{3}{7}$ of 42 = _____

Find the perimeter of each square or rectangle.

10.

32 yd

81 yd

P = _____

11.

28'

28'

P = _____

12. Thirty-five people tried out for the play. Three-fifths of them were chosen to be in the play. How many were chosen?

13. What is the perimeter of a square patio that is 10 feet on a side?

14. A bench was built along 1/5 of the perimeter of the patio in #13. How long is the bench?

15. Chance had $16.45 in his wallet. He lost $5.30. How much money does he have left?

3

Build, then shade, and write your answer in words and symbols.

1.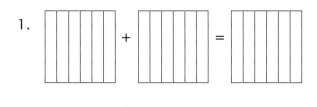

    ____ + ____ = ____

    one-sixth plus four-sixths equals _____

2.

    ____ - ____ = ____

    three-fourths minus one-fourth equals _____

Add or subtract.

3. $\dfrac{3}{5} + \dfrac{2}{5} =$ ___

4. $\dfrac{4}{6} + \dfrac{1}{6} =$ ___

5. $\dfrac{2}{10} + \dfrac{6}{10} =$ ___

6. $\dfrac{5}{8} - \dfrac{3}{8} =$ ___

7. $\dfrac{8}{9} - \dfrac{2}{9} =$ ___

8. $\dfrac{2}{3} - \dfrac{1}{3} =$ ___

Solve.

9. $\dfrac{1}{8}$ of 24 = _____

10. $\dfrac{4}{5}$ of 10 = _____

11. $\dfrac{4}{10}$ of 20 = _____

Find the perimeter of each shape.

12.
98'
59'
54'

13.
8 yd
12 yd

14.
16"
16"

P = _____

P = _____

P = _____

15. Ben ate 1/6 of an apple pie and 2/6 of a blueberry pie. How much pie did he eat in all?

16. Katherine has finished 2/3 of the job her mother gave her to do. What part of the job is left to to be done?

17. Elizabeth plans to spend 3/4 of the year at school in New Hampshire. How many months is that?

18. Ruth and Clara had a box of one dozen donuts. Ruth ate 1/6 of them and Clara ate 2/6 of them. What part of the donuts has been eaten?
How many donuts is that?

19. What part of the donuts in #18 is left to eat?
How many donuts is that?

20. Caleb received $15.50, $5.25, and $10.00 as gifts. How much money did he receive in all?

Build the equivalent fractions. Draw lines and fill in the blanks to show what you have built.
Write the fractions with numbers and with words.

1.

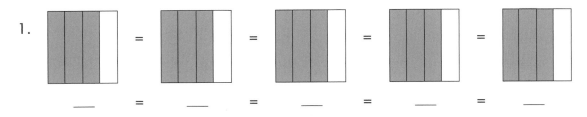

$\underline{\quad}$ = $\underline{\quad}$ = $\underline{\quad}$ = $\underline{\quad}$ = $\underline{\quad}$

_____ _____ _____ _____ _____

Fill in the missing numbers to make equivalent fractions.

2. $\dfrac{1}{5} = \dfrac{\quad}{\quad} = \dfrac{\quad}{15} = \dfrac{4}{\quad}$

3. $\dfrac{5}{6} = \dfrac{\quad}{\quad} = \dfrac{15}{\quad} = \dfrac{\quad}{24}$

4. $\dfrac{3}{7} = \dfrac{\quad}{\quad} = \dfrac{\quad}{\quad} = \dfrac{12}{28}$

5. $\dfrac{1}{2} = \dfrac{\quad}{\quad} = \dfrac{3}{\quad} = \dfrac{\quad}{8}$

Add or subtract.

6. $\dfrac{4}{5} - \dfrac{1}{5} = \dfrac{\quad}{\quad}$

7. $\dfrac{5}{8} + \dfrac{2}{8} = \dfrac{\quad}{\quad}$

8. $\dfrac{5}{9} - \dfrac{4}{9} = \dfrac{\quad}{\quad}$

Solve.

9. $\dfrac{3}{8}$ of 48 = $\underline{\quad}$

10. $\dfrac{1}{6}$ of 18 = $\underline{\quad}$

11. $\dfrac{2}{3}$ of 27 = $\underline{\quad}$

Multiply.

12.   2 2
      × 3 1

13.   5 4
      × 1 1

14.   3 2 2
      × 2 2

15.   Lauren has colored one-half of the pages in her coloring book. How many fourths has she colored?

16.   If Lauren's book (#15) has 20 pages, how many has she colored?

17.   If Rebekah is 13 years old, how many months has she lived?

18.   Each side of a square measures 11 feet. What is the perimeter of the square?

19.   Annika read 2/7 of her book before lunch and 3/7 of it after lunch. What part of her book is left to read?

20.   Annika's book has 56 pages (#19). How many pages are left to read?

Use the overlays to find equivalent fractions, then add or subtract.

1. $\dfrac{3}{6} + \dfrac{2}{5} =$ ———

2. $\dfrac{2}{4} + \dfrac{1}{3} =$ ———

3. $\dfrac{1}{5} + \dfrac{1}{6} =$ ———

4. $\dfrac{1}{4} - \dfrac{1}{5} =$ ———

5. $\dfrac{5}{6} - \dfrac{2}{5} =$ ———

6. $\dfrac{1}{2} - \dfrac{1}{4} =$ ———

Fill in the missing numbers in the numerators or denominators to make equivalent fractions.

7. $\dfrac{1}{7} = \dfrac{}{} = \dfrac{}{21} = \dfrac{4}{}$

8. $\dfrac{2}{3} = \dfrac{}{} = \dfrac{}{} = \dfrac{8}{12}$

Solve.

9. $\dfrac{5}{9}$ of 36 = ———

10. $\dfrac{3}{10}$ of 90 = ———

11. $\dfrac{4}{6}$ of 48 = ———

Round to the nearest ten.

12. 85 → _____

13. 13 → _____

Round to the nearest hundred.

14. 250 → _____

15. 175 → _____

16. One-sixth of the soldiers rode in tanks and one-third of them rode in trucks. The rest of them had to walk. What part of the soldiers was able to ride? (Use the overlays.)

17. What part of the soldiers in #16 had to walk? (18/18 is all the soldiers.)

18. Zarah bought 2/3 of a pound of nuts. She used 1/2 of a pound of nuts in the cake she baked. What part of a pound of nuts is left over? (Use the overlays.)

19. Sharon makes and sells bread. If each loaf takes three cups of flour, how much flour will she need for 122 loaves?

20. The sides of a triangle are 7 feet, 24 feet, and 25 feet. What is the perimeter of the triangle?

6

Add or subtract using the rule of four.

1. $\dfrac{1}{3} + \dfrac{2}{5} = $ ———

2. $\dfrac{3}{4} + \dfrac{1}{6} = $ ———

3. $\dfrac{3}{7} + \dfrac{1}{10} = $ ———

4. $\dfrac{1}{2} - \dfrac{1}{6} = $ ———

5. $\dfrac{5}{8} - \dfrac{1}{3} = $ ———

6. $\dfrac{5}{6} - \dfrac{3}{4} = $ ———

Fill in the missing numbers in the numerators or denominators to make equivalent fractions.

7. $\dfrac{7}{8} = \dfrac{}{} = \dfrac{}{24} = \dfrac{28}{}$

8. $\dfrac{1}{3} = \dfrac{}{} = \dfrac{}{} = \dfrac{}{12}$

Solve.

9. $\dfrac{1}{10}$ of 50 = ____

10. $\dfrac{3}{4}$ of 16 = ____

11. $\dfrac{1}{2}$ of 20 = ____

Estimate, then multiply to find the exact answer.

12. $\begin{array}{r} 2\,6 \\ \times\,3\,3 \\ \hline \end{array}$ →

13. $\begin{array}{r} 7\,5 \\ \times\,8\,3 \\ \hline \end{array}$ →

14. $\begin{array}{r} 6\,1 \\ \times\,1\,7 \\ \hline \end{array}$ →

15.   Sixteen boxes of books arrived in the mail. If each box held one dozen books, how many books did I receive?

16.   One-half of Hannibal's elephants died as they struggled over the Alps. Another one-fifth did not survive the march down Italy. What part of his elephants did Hannibal lose?

17.   Mom used 4/9 of the apples to make applesauce and 3/6 of of them to make a pie. What part of the apples has been used?

18.   Gavin saw 3/4 of a pizza on the counter. By the time he was finished eating, there was only 1/8 of a pizza left. What part of a pizza did Gavin eat?

19.   Round 472 to the nearest hundred.

20.   Vontoria took 7/10 of an hour to do her math today. How many minutes did she spend on math today?

Use the rule of four to compare the fractions and write the correct symbol in the oval.

1. $\dfrac{1}{4}$ ◯ $\dfrac{3}{7}$

2. $\dfrac{3}{8}$ ◯ $\dfrac{1}{2}$

3. $\dfrac{4}{5}$ ◯ $\dfrac{2}{9}$

4. $\dfrac{6}{11}$ ◯ $\dfrac{2}{3}$

5. $\dfrac{5}{9}$ ◯ $\dfrac{6}{7}$

6. $\dfrac{3}{4}$ ◯ $\dfrac{6}{8}$

Add or subtract.

7. $\dfrac{1}{8} + \dfrac{4}{9} = $ _____

8. $\dfrac{2}{3} - \dfrac{1}{5} = $ _____

9. $\dfrac{4}{5} + \dfrac{1}{6} = $ _____

Fill in the missing numbers in the numerators or denominators to make equivalent fractions.

10. $\dfrac{2}{5} = $ _____ $ = $ _____ $ = $ _____

Divide and write your remainder as a fraction.

11. $2\overline{)19}$

12. $6\overline{)51}$

13. $5\overline{)39}$

Estimate, then multiply to find the exact answer.

14. $\begin{array}{r} 3\ 9 \\ \times\ 2\ 4 \\ \hline \end{array}$ →

15. $\begin{array}{r} 7\ 2 \\ \times\ 1\ 5 \\ \hline \end{array}$ →

16. $\begin{array}{r} 6\ 8 \\ \times\ 4\ 3 \\ \hline \end{array}$ →

17.   Five-eighths of the trees in my yard are maples. If there are 16 trees in my yard, how many are maple trees?

18.   Christa ate 1/5 of the chocolates in the box and Douglas ate 3/10 of them. Write a comparison showing who ate the most chocolates.

19.   If there were 20 chocolates in the box in #18, how many chocolates did each person eat? Write another comparison using the actual number of chocolates eaten. Does it agree with the comparison you wrote for #18?

20.   Yesterday we got 2/10 of an inch of rain. Today we had 8/10 of an inch of rain. How much more rain did we get today?

8

Use whichever method you prefer to add the fractions.

1. $\dfrac{1}{2} + \dfrac{3}{7} + \dfrac{1}{3} = $ ——

2. $\dfrac{4}{6} + \dfrac{1}{4} + \dfrac{2}{3} = $ ——

3. $\dfrac{3}{4} + \dfrac{3}{8} + \dfrac{1}{2} = $ ——

4. $\dfrac{3}{4} + \dfrac{1}{6} + \dfrac{1}{5} = $ ——

5. $\dfrac{2}{8} + \dfrac{1}{2} + \dfrac{2}{5} = $ ——

6. $\dfrac{1}{2} + \dfrac{2}{3} + \dfrac{5}{6} = $ ——

Use the rule of four to compare the fractions and write the correct symbol in the oval.

7. $\dfrac{2}{4}$ ◯ $\dfrac{4}{8}$

8. $\dfrac{3}{7}$ ◯ $\dfrac{3}{8}$

9. $\dfrac{2}{5}$ ◯ $\dfrac{3}{4}$

Fill in the missing numbers in the numerators or denominators to make equivalent fractions.

10. $\dfrac{3}{9} = $ —— $ = $ —— $ = \dfrac{}{36}$

11. —— $ = \dfrac{4}{6} = $ —— $ = \dfrac{8}{}$

Divide and write your remainder as a fraction.

12. $3\overline{)298}$ 13. $7\overline{)156}$ 14. $4\overline{)465}$

15. Brayden planted tomatoes in his garden. One-fourth of the plants got eaten by pests. One-half of them died because Brayden forgot to water them, and one-eighth of the plants was killed when the dog dug a hole in the garden. What part of Brayden's tomato plants died or were destroyed?

16. What part of Brayden's plants is still alive? (#15) If he started with 32 plants, how many plants are left?

17. Each of Brayden's surviving plants (#16) produced 65 tomatoes during the summer. How many tomatoes did he get in all?

18. Mr. Knowles divided $217 evenly among his five children. How many dollars did each receive? (Write your remainder as a fraction.)

19. What is 61 rounded to the nearest ten?

20. What is 829 rounded to the nearest hundred?

Solve.

1. $\frac{2}{3}$ of 12 = _____

2. $\frac{3}{5}$ of 10 = _____

3. $\frac{3}{4}$ of 24 = _____

4. $\frac{1}{3}$ of 15 = _____

5. $\frac{4}{5}$ of 25 = _____

6. $\frac{5}{7}$ of 14 = _____

Fill in the missing numbers in the numerators or denominators to make equivalent fractions.

7. $\frac{2}{6} = \dfrac{}{} = \dfrac{}{} = \dfrac{}{24}$

8. $\dfrac{}{} = \dfrac{10}{16} = \dfrac{}{} = \dfrac{20}{}$

Add or subtract using the rule of four when needed.

9. $\dfrac{1}{3} + \dfrac{1}{3} =$ ———

10. $\dfrac{1}{4} + \dfrac{3}{6} =$ ———

11. $\dfrac{1}{5} + \dfrac{5}{6} =$ ———

12. $\dfrac{2}{3} - \dfrac{1}{5} =$ ———

13. $\dfrac{7}{10} - \dfrac{3}{10} =$ ——

14. $\dfrac{4}{7} - \dfrac{3}{8} =$ ———

Use the rule of four to compare the fractions and write the correct symbol in the oval.

15. $\dfrac{3}{5} \bigcirc \dfrac{5}{7}$

16. $\dfrac{3}{6} \bigcirc \dfrac{4}{8}$

17. $\dfrac{2}{3} \bigcirc \dfrac{4}{8}$

Use whichever method you prefer to add the fractions.

18. $\dfrac{3}{8} + \dfrac{1}{10} + \dfrac{2}{5} =$ ———

19. $\dfrac{3}{4} + \dfrac{7}{8} + \dfrac{1}{2} =$ ———

Find the perimeter of each shape.

20.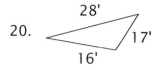

28'

17'

16'

P = _____

21.

23 yd

46 yd

P = _____

22.

52"

52"

P = _____

Round to the nearest ten.

23. 91 → _____

24. 45 → _____

Round to the nearest hundred.

25. 604 → _____

26. 517 → _____

Divide and write your remainder as a fraction.

27. 5⟌3 5 8   28. 8⟌5 4 1   29. 6⟌1 8 9

Multiply.

30.   6 7
      × 1 8

31.   3 4
      × 2 6

32.   2 2 4
      × 2 2

33. Sophia used 2/5 of her birthday money for school supplies and 1/3 of it for gifts. What part of her birthday money has she spent so far?

34. During our last snow storm, Drumore received 5/6 of a foot and Quarryville received 2/3 of a foot. Tell which town received more snow, then find the difference in their amounts of snowfall.

35. Charles has a rectangular room that measures 10 feet by 11 feet. One-seventh of the perimeter of the room is taken up by bookshelves. How many feet of bookshelves does Charles have in his room?

9

Multiply (fraction of a fraction).

1. $\frac{1}{4}$ of $\frac{1}{3}$ = ——

2. $\frac{3}{4} \times \frac{2}{5}$ = ——

3. $\frac{2}{7} \times \frac{1}{4}$ = ——

4. $\frac{2}{5}$ of $\frac{3}{7}$ = ——

5. $\frac{3}{5} \times \frac{1}{6}$ = ——

6. $\frac{2}{4} \times \frac{1}{3}$ = ——

Add or subtract.

7. $\frac{3}{11} + \frac{1}{4}$ = ——

8. $\frac{4}{5} - \frac{1}{6}$ = ——

9. $\frac{1}{7} + \frac{2}{3} + \frac{1}{2}$ = ——

Use the rule of four to make denominators the same, then compare the fractions.

10. $\dfrac{2}{3} \bigcirc \dfrac{5}{8}$ 11. $\dfrac{3}{4} \bigcirc \dfrac{9}{12}$ 12. $\dfrac{5}{6} \bigcirc \dfrac{7}{9}$

Estimate, then multiply to find the exact answer.

13. $\begin{array}{r} 612 \rightarrow \\ \times\ 54 \\ \hline \end{array}$ 14. $\begin{array}{r} 124 \rightarrow \\ \times\ 36 \\ \hline \end{array}$ 15. $\begin{array}{r} 957 \rightarrow \\ \times\ 13 \\ \hline \end{array}$

16. One-third of the customers at the ice cream store bought vanilla and two-fifths of them bought chocolate. What part of the customers bought vanilla or chocolate?

17. What is the perimeter of a triangle whose sides measure 10 feet, 12 feet, and 12 feet?

18. Three-eighths of the guests at the picnic ate hamburgers. One-half of those people had mustard on their hamburgers. What part of the people at the picnic had hamburgers with mustard?

19. If there were 48 people at the picnic (#18), how many had hamburgers? How many had hamburgers with mustard?

20. Each of the 48 people at the picnic contributed $15 for food and other expenses. How much money was collected?

Divide. If the denominators are different, use the rule of four first to make them the same.

1. $\dfrac{2}{5} \div \dfrac{1}{5} =$ ——

2. $\dfrac{5}{8} \div \dfrac{2}{3} =$ ——

3. $\dfrac{1}{2} \div \dfrac{1}{4} =$ ——

4. $\dfrac{1}{4} \div \dfrac{1}{3} =$ ——

5. $\dfrac{3}{5} \div \dfrac{2}{9} =$ ——

6. $\dfrac{4}{5} \div \dfrac{2}{3} =$ ——

Multiply.

7. $\dfrac{3}{4} \times \dfrac{1}{4} =$ ——

8. $\dfrac{2}{3} \times \dfrac{1}{5} =$ ——

9. $\dfrac{2}{9} \times \dfrac{1}{2} =$ ——

Add or subtract.

10. $\dfrac{4}{7} + \dfrac{3}{8} =$ ——

11. $\dfrac{5}{9} - \dfrac{1}{4} =$ ——

12. $\dfrac{2}{5} + \dfrac{3}{10} + \dfrac{1}{2} =$ ——

Round and estimate. Don't worry about any remainders.

13. $38\overline{)541}$ →

14. $26\overline{)876}$ →

15. $55\overline{)605}$ →

16. If a board is 1/12 of a foot thick, how many pieces are in a pile 1/2 foot high?

17. Brooke has a piece of ribbon that is 7/8 of a yard long. Into how many 1/8-yard-long pieces can she cut it?

18. Jillian walked 3/10 of a mile and ran 3/5 of a mile. Then she roller-skated 1/10 of a mile. How far did Jillian travel?

19. The first time Ned checked, he saw that 7/8 of the pizza was left in the pan. Later he noticed that there was only 1/4 of the pizza left in the pan. What part of a pizza was taken since he checked the first time?

20. Since there are 24 hours in a day, how many hours are there in 365 days (one year)?

Use the divisibility rules to answer the questions. Write yes or no in the blanks.

1. Is 17 divisible by 2? ____

2. Is 105 divisible by 5? ____

3. Is 23 divisible by 3? ____

4. Is 288 divisible by 9? ____

List all the factors for each number, then underline the common factors. Write the greatest common factor (GCF) for each pair of numbers in the blank.

5.   8:

24:

6. 10:

20:

The GCF of 8 and 24 is ____

The GCF of 10 and 20 is ____

7. 39:

15:

8. 14:

28:

The GCF of 39 and 15 is ____

The GCF of 14 and 28 is ____

Multiply.

9. $\dfrac{1}{8} \times \dfrac{2}{7} =$ ——

10. $\dfrac{2}{9} \times \dfrac{4}{5} =$ ——

11. $\dfrac{3}{6} \times \dfrac{1}{6} =$ ——

Divide using the rule of four.

12. $\dfrac{1}{8} \div \dfrac{1}{2} = $ ____

13. $\dfrac{2}{3} \div \dfrac{1}{6} = $ ____

14. $\dfrac{5}{6} \div \dfrac{3}{10} = $ ____

Use the rule of four to make denominators the same, then compare the fractions.

15. $\dfrac{3}{4} \bigcirc \dfrac{9}{12}$

16. $\dfrac{3}{8} \bigcirc \dfrac{4}{10}$

17. $\dfrac{1}{2} \bigcirc \dfrac{5}{9}$

Divide. Write any remainders as fractions.

18. $2\,6\,\overline{|\,6\,7\,3}$

19. $8\,2\,\overline{|\,3\,9\,0}$

20. $5\,1\,\overline{|\,7\,6\,8}$

Find the GCF of the numerator and denominator and use it to reduce each fraction.

1. $\dfrac{8}{10} \div \underline{} = \underline{}$

2. $\dfrac{12}{20} \div \underline{} = \underline{}$

3. $\dfrac{27}{30} \div \underline{} = \underline{}$

4. $\dfrac{20}{30} \div \underline{} = \underline{}$

5. $\dfrac{15}{25} \div \underline{} = \underline{}$

6. $\dfrac{36}{48} \div \underline{} = \underline{}$

Use the divisibility rules to answer the questions. Write yes or no in the blanks.

7. Is 18 divisible by 3? _____

8. Is 556 divisible by 5? _____

9. Is 41 divisible by 2? _____

10. Is 372 divisible by 9? _____

Follow the signs, then use the GCF to reduce your answers.

11. $\dfrac{3}{6} + \dfrac{1}{4} = \underline{} = \underline{}$

12. $\dfrac{2}{9} + \dfrac{1}{6} = \underline{} = \underline{}$

13. $\dfrac{2}{3} \times \dfrac{5}{6} =$ —— = ——   14. $\dfrac{5}{8} \times \dfrac{1}{10} =$ —— = ——

15. $\dfrac{1}{5} \div \dfrac{8}{10} =$ —— = ——   16. $\dfrac{4}{7} \div \dfrac{2}{3} =$ —— = ——

Multiply. Use lined paper if you wish.

17.
```
    5 7 3
  × 6 1 2
```

18.
```
   3,6 8 2
  ×   6 9 4
```

19. Three-eighths of the wall was left to be painted. Arik painted one-half of what was left to be done. What part of the whole wall did Arik paint?

20. Colleen walked 6/8 of a mile. She divided her walk into 1/4-mile sections so she could stop for stretching exercises. Into how many parts was her walk divided?

Find the prime factors using whatever method you wish.

1. $50 =$ _____

2. $48 =$ _____

3. $63 =$ _____

4. $99 =$ _____

Reduce the fractions using prime factors.

5. $\dfrac{20}{30} =$ ——————————— $=$ ———

6. $\dfrac{63}{81} =$ ——————— $=$ ——————— $=$ ———

7. $\dfrac{36}{42} =$ ——————— $=$ ——————— $=$ ———

Follow the signs, then reduce using GCF or prime factors.

8. $\dfrac{1}{6} + \dfrac{5}{12} =$ —— $=$ ——

9. $\dfrac{5}{7} \div \dfrac{5}{6} =$ —— $=$ ——

10. $\dfrac{3}{4} \times \dfrac{1}{3} =$ —— $=$ ——

Use the rule of four to make denominators the same, then compare the fractions.

11. $\dfrac{4}{7} \bigcirc \dfrac{3}{8}$ 12. $\dfrac{5}{10} \bigcirc \dfrac{6}{11}$ 13. $\dfrac{1}{4} \bigcirc \dfrac{2}{15}$

Divide. Write any remainders as fractions. Use lined paper if you wish.

14. $8\,3\,\overline{\smash{)}\,5{,}3\,1\,1}$ 15. $3\,6\,\overline{\smash{)}\,8{,}8\,5\,6}$ 16. $5\,1\,2\,\overline{\smash{)}\,6{,}5\,3\,2}$

17. Paul planted 25 strawberry plants. If each plant yielded 17 quarts of berries this summer, how many quarts of berries did Paul get in all?

18. Becky drove for 1/7 of her trip and flew for 7/10 of it. She walked for the rest of the trip. For what part of the trip was Becky walking?

19. One-half of the customers at the store were shopping for school supplies. One-third of those customers bought backpacks. What part of the total customers bought backpacks?

20. Two-sevenths of the total customers in #19 bought dictionaries. Which group was larger, those who bought backpacks or those who bought dictionaries?

Write the length of each line beneath the ruler. Reduce if possible. If not, leave the second blank empty.

1. 0"   1"

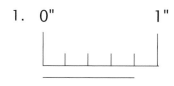

_____

____ " = ____ "

2. 0"   1"

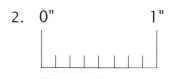

_____

____ " = ____ "

3. 0"   1"

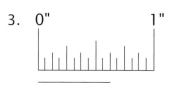

_____

____ " = ____ "

4. 0"   1"

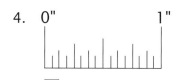

_____

____ " = ____ "

5. 0"   1"

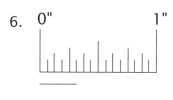

_____

____ " = ____ "

6. 0"   1"

_____

____ " = ____ "

Find the prime factors using whatever method you wish.

7. 24 = _____   8. 76 = _____

9. 48 = _____

Reduce the fractions using prime factors.

10. $\dfrac{32}{54}$ = ——————————— = —————— = ——

Find the area of each rectangle.

11. [rectangle] 15"
32"

12. [rectangle] 4'
10'

13. [rectangle] 24 yd
49 yd

A = _____   A = _____   A = _____

14. Is 168 divisible by nine?

15. What is the GCF of 16 and 44?

16. Two-fifths of the days in April were cloudy. It rained on 3/4 of the cloudy days. What part of the days in April was rainy?

17. A grocer bought 650 pounds of potatoes from a local farmer. If he put them in 10-pound bags for sale, how many bags did he use?

18. Wade wants to put a fence around his garden. Does he need to find the area or the perimeter of the garden?

Change each mixed number to an improper fraction.

1. $1\frac{2}{9} = \frac{}{9} + \frac{}{9} = $ ———

2. $3\frac{3}{5} = \frac{}{5} + \frac{}{5} + \frac{}{5} + \frac{}{5} = $ ———

3. $1\frac{1}{3} = \frac{}{3} + \frac{}{3} = $ ———

4. $2\frac{1}{4} = \frac{}{4} + \frac{}{4} + \frac{}{4} = $ ———

Change each improper fraction to a mixed number.

5. $\frac{7}{4} = \frac{}{4} + \frac{}{4} = $ ———

6. $\frac{9}{2} = \frac{}{2} + \frac{}{2} + \frac{}{2} + \frac{}{2} + \frac{}{2} = $ ———

7. $\frac{13}{5} = \frac{}{5} + \frac{}{5} + \frac{}{5} = $ ———

8. $\frac{23}{6} = \frac{}{6} + \frac{}{6} + \frac{}{6} + \frac{}{6} = $ ———

Write the length of each line beneath the ruler. Reduce if possible. If not, leave the second blank empty.

9. 0"   1"

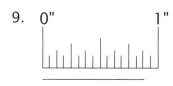

——————

——— " = ——— "

10. 0"   1"

——————

——— " = ——— "

Reduce the fractions using any method you wish.

11. $\frac{3}{9} = $ ———

12. $\frac{20}{28} = $ ———

13. $\frac{18}{32} = $ ———

Find the area of a square with the given side.

14. S = 16 inches

A = _____

15. S = 3 miles

A = _____

Solve.

16. 8^2 = _____   17. 6^2 = _____

18. Is 213 divisible by three?

19. What is the GCF of 15 and 30?

20. What are the prime factors of 45?

Write the length of the line.

1. _____ "

2. _____ "

Change each mixed number to an improper fraction.

3. $1\dfrac{1}{8} = \dfrac{}{8} + \dfrac{}{8} = $ _____

4. $2\dfrac{5}{6} = \dfrac{}{6} + \dfrac{}{6} + \dfrac{}{6} = $ _____

Change each improper fraction to a mixed number.

5. $\dfrac{9}{4} = \dfrac{}{4} + \dfrac{}{4} + \dfrac{}{4} = $ _____

6. $\dfrac{24}{7} = \dfrac{}{7} + \dfrac{}{7} + \dfrac{}{7} + \dfrac{}{7} = $ _____

Follow the signs. Reduce if possible and write any improper fractions as mixed numbers.

7. $\dfrac{3}{7} + \dfrac{2}{9} =$ _____

8. $\dfrac{1}{8} + \dfrac{1}{6} =$ _____

9. $\dfrac{1}{3} - \dfrac{2}{8} =$ _____

10. $\dfrac{5}{6} \times \dfrac{11}{12} =$ _____

11. $\dfrac{1}{2} \div \dfrac{3}{10} =$ _____

12. $\dfrac{2}{3} \times \dfrac{7}{20} =$ _____

Find the area and perimeter of each figure.

13. A = _____

14. P = _____

15. A = _____

16. P = _____

17. A = _____

18. P = _____

19. Solve: 16^2

20. What is the GCF of 36 and 42?

Divide. Reduce answers when possible and write any improper fractions as mixed numbers.

1. $\dfrac{1}{2} \div \dfrac{5}{9} =$ _____

2. $\dfrac{3}{7} \div \dfrac{5}{6} =$ _____

3. $\dfrac{4}{7} \div \dfrac{5}{14} =$ _____

4. $\dfrac{7}{9} \div \dfrac{2}{3} =$ _____

5. $\dfrac{3}{4} \div \dfrac{1}{8} =$ _____

6. $\dfrac{1}{4} \div \dfrac{3}{5} =$ _____

Multiply. Reduce answers when possible.

7. $\dfrac{1}{4} \times \dfrac{1}{3} =$ ___

8. $\dfrac{4}{5} \times \dfrac{7}{10} =$ ___

9. $\dfrac{2}{5} \times \dfrac{1}{2} =$ ___

10. $\dfrac{3}{4} \times \dfrac{1}{4} =$ ___

11. $\dfrac{3}{7} \times \dfrac{2}{9} =$ ___

12. $\dfrac{1}{8} \times \dfrac{1}{6} =$ ___

Use the divisibility rules to answer the questions. Write yes or no in the blanks.

13. Is 18 divisible by 2? ____

14. Is 552 divisible by 5? ____

15. Is 102 divisible by 3? ____

16. Is 962 divisible by 9? ____

17. What is the GCF of 28 and 54?

18. What are the prime factors of 60?

Reduce the fractions using any method you wish.

19. $\dfrac{4}{12} =$ ——

20. $\dfrac{15}{20} =$ ——

21. $\dfrac{18}{42} =$ ——

Change each mixed number to an improper fraction.

22. $1\dfrac{3}{4} = \dfrac{}{4} + \dfrac{}{4} =$ ——

23. $2\dfrac{7}{9} = \dfrac{}{9} + \dfrac{}{9} + \dfrac{}{9} =$ ——

Change each improper fraction to a mixed number.

24. $\dfrac{13}{5} = \dfrac{}{5} + \dfrac{}{5} + \dfrac{}{5} = $ ———

25. $\dfrac{10}{3} = \dfrac{}{3} + \dfrac{}{3} + \dfrac{}{3} + \dfrac{}{3} = $ ——

Write the length of the line.

26. _____ "

27. _____ "

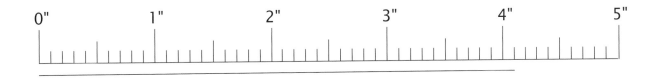

28. _____ "

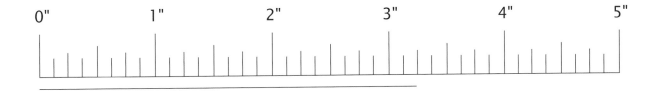

Find the area of each figure.

 3'

7'

 12"

12"

 10'

6'

8'

29. A = _____

30. A = _____

31. A = _____

32. Solve: 9^2

33. Marcy had 3/4 of her birthday cake left over. She wanted to give each of her guests 1/16 of a whole cake. How many people can she serve?

34. Before Marcy could serve her guests, her dog came along and ate one-half of the cake that was left over (#33). What part of a whole cake did the dog eat?

Add or subtract. Continue to estimate your answers mentally.

1.　　$2\frac{1}{10}$

　　$+\ 6\frac{6}{10}$

2.　　$4\frac{3}{5}$

　　$-\ 3\frac{1}{5}$

3.　　$5\frac{2}{7}$

　　$+\ 1\frac{4}{7}$

Change each mixed number to an improper fraction. Write out the steps only if you need to.

4.　$6\frac{1}{2}$ = ——

5.　$5\frac{2}{3}$ = ——

6.　$1\frac{9}{10}$ = ——

Change each improper fraction to a mixed number.

7.　$\frac{21}{4}$ = _____

8.　$\frac{11}{5}$ = _____

9.　$\frac{15}{8}$ = _____

Follow the signs. Reduce if possible.

10.　$\frac{5}{6} \times \frac{1}{3}$ = _____

11.　$\frac{1}{10} \div \frac{3}{10}$ = _____

12.　$\frac{8}{9} \times \frac{5}{6}$ = _____

43

Find the volume of each cube.

13. V = _____   14. V = _____   15. V = _____

16. Andrew built a vacation house that was a cube measuring 17 feet by 17 feet by 17 feet. What was the volume of his new house?

17. What are the prime factors of 95?

18. Briana went shopping and bought 4 3/10 pounds of hamburger and 2 6/10 pounds of steak. How many pounds of meat did she buy?

19. A square is 3/4 of a mile on each side. What is the area of the square?

20. One hundred twenty people came to watch the ball game at the local park. Five-eighths of them were parents of the players. How many people were parents of the players?

18

Add the mixed numbers.

1. $6\frac{4}{7}$

    $+\ 7\frac{5}{7}$
    _____

2. $13\frac{4}{9}$

    $+\ 5\frac{8}{9}$
    _____

3. $21\frac{11}{12}$

    $+\ 3\frac{1}{12}$
    _____

4. $4\frac{3}{5}$

    $+\ 8\frac{4}{5}$
    _____

5. $9\frac{2}{3}$

    $+\ 8\frac{2}{3}$
    _____

6. $6\frac{7}{8}$

    $+\ 7\frac{3}{8}$
    _____

Follow the signs.

7. $\frac{3}{4} - \frac{1}{5} = $ _____

8. $\frac{1}{10} + \frac{1}{5} = $ _____

9. $\frac{7}{9} - \frac{1}{2} = $ _____

10. $\frac{3}{10} \div \frac{1}{10}$ _____

11. $\frac{5}{6} \times \frac{1}{3} = $ _____

12. $\frac{3}{4} \div \frac{1}{3} = $ _____

45

Solve. Remember that the times sign means the same as "of."

13. $\dfrac{3}{8} \times 64 =$ _____

14. $\dfrac{1}{6} \times 42 =$ _____

15. $\dfrac{7}{10} \times 80 =$ _____

16. Kara built a rectangular fish pond in her garden. It was eight feet long, five feet wide and two feet deep. How many cubic feet of water would be required to fill it to the top?

17. At 56 pounds per cubic foot, how much will the water in #16 weigh?

18. What is the GCF of 42 and 56?

19. What are the prime factors of 18?

20. Aaron ran 2 4/5 of a mile. Tony ran as far as Aaron had, then he ran 1 3/5 miles farther. How many miles did Tony run?

Add or subtract the mixed numbers, regrouping as needed.

1. $\quad 2\dfrac{1}{8}$

$\quad -1\dfrac{5}{8}$

2. $\quad 5\dfrac{2}{7}$

$\quad -\dfrac{3}{7}$

3. $\quad 3$

$\quad -2\dfrac{1}{4}$

4. $\quad 2\dfrac{5}{8}$

$\quad +2\dfrac{4}{8}$

5. $\quad 7\dfrac{3}{5}$

$\quad +4\dfrac{4}{5}$

6. $\quad 8\dfrac{3}{6}$

$\quad +9\dfrac{3}{6}$

Follow the signs.

7. $\dfrac{4}{5} \div \dfrac{2}{3} = $ _____

8. $\dfrac{2}{11} \times \dfrac{1}{5} = $ _____

9. $\dfrac{1}{2} + \dfrac{2}{3} + \dfrac{5}{6} = $ _____

Use the rule of four to make denominators the same, then compare the fractions.

10. $\frac{1}{4}$ ◯ $\frac{2}{5}$ 　　　　　11. $\frac{3}{7}$ ◯ $\frac{1}{8}$

12. $\frac{5}{12}$ ◯ $\frac{4}{10}$

Fill in the blanks.

13. 99 ft = ____ yd 　　　　　14. 26 yd = ____ ft

15. 9 yd = ____ ft

16. A rectangle is 2/3 of a mile long and 1/3 of a mile wide. What is the area of the rectangle?

17. What is the perimeter of the rectangle in #16?

18. A restaurant sold 9 5/8 pies on Monday and 6 4/8 pies on Tuesday. How many pies were sold during those two days?

19. Melissa bought 15 1/8 yards of fabric for curtains. If she uses only 11 3/8 yards, how much fabric will she have left over?

20. If Mark is six feet tall, how many yards tall is he?

Subtract using the same difference theorem.

1. $8 \ \dfrac{5}{8}$ +   =

    $- \ 3 \ \dfrac{7}{8}$ +   =

2. 14 +   =

    $- \ 10 \ \dfrac{1}{3}$ +   =

Add or subtract. Use whichever method you prefer to subtract.

3. $2\dfrac{3}{4}$

    $+ \ 3\dfrac{3}{4}$

4. $7\dfrac{1}{3}$

    $- \ 3\dfrac{2}{3}$

5. $4\dfrac{3}{5}$

    $+ \ 1\dfrac{4}{5}$

Add or subtract.

6. $\dfrac{2}{3} + \dfrac{3}{4} =$ _____

7. $\dfrac{3}{4} + \dfrac{5}{6} =$ _____

8. $\dfrac{1}{2} + \dfrac{4}{7} =$ _____

9. $\dfrac{7}{8} - \dfrac{1}{3} =$ _____

10. $\dfrac{5}{12} - \dfrac{2}{10} =$ _____

11. $\dfrac{1}{6} - \dfrac{1}{7} =$ _____

Fill in the blanks.

12. 32 qt = ____ pt

13. 34 pt = ____ qt

14. 7 yd = ____ ft

15. A farmer got 9 3/4 tons of hay the first time he cut the field and 6 3/4 tons the second time he cut it. What was the total amount of hay he harvested?

16. Luke had 4 2/5 gallons of gasoline in his tank when he started his trip. If he used 1 4/5 gallons, how much was left in the tank after the trip?

17. Is 616 divisible by 3?

18. A cube measures nine feet on a side. Give the volume in cubic yards. (Convert feet to yards first.)

19. A square measures 2 1/4 feet on each side. What is the perimeter of the square?

20. Solve: 15^2

Add the mixed numbers

1. $9\frac{1}{3}$

 $+ 6\frac{1}{4}$

2. $4\frac{2}{3}$

 $+ 1\frac{3}{5}$

3. $9\frac{3}{5}$

 $+ 2\frac{7}{10}$

4. $12\frac{5}{11}$

 $+ 4\frac{5}{8}$

Subtract. Use whichever method you prefer.

5. $5\frac{1}{5}$

 $- 2\frac{3}{5}$

6. $15\frac{7}{9}$

 $- 6\frac{2}{9}$

7. 7

 $- 3\frac{1}{5}$

Divide or multiply.

8. $\frac{1}{2} \div \frac{1}{4} =$ _____

9. $\frac{8}{12} \div \frac{1}{3} =$ _____

10. $\dfrac{3}{5} \div \dfrac{1}{10} =$ _____

11. $\dfrac{4}{6} \times \dfrac{3}{5} =$ _____

12. $\dfrac{1}{2} \times \dfrac{1}{3} =$ _____

13. $\dfrac{3}{4} \times \dfrac{1}{6} =$ _____

Fill in the blanks.

14. 81 ft = _____ yd

15. 54 pt = _____ qt

16. 18 qt = _____ pt

17. How many yards are there in 92 feet? Write your remainder as a fraction.

18. Duncan drove 20 1/2 miles, then walked 2 3/4 miles. How far did Duncan travel in all?

19. Each side of a square is 2/3 of a foot long. Find the area and the perimeter of the square.

20. Mental Math! 3 times 4, times 2, plus 1, divided by 5, equals _____ .

Solve.

1. $4\dfrac{1}{4}$
$-2\dfrac{3}{4}$
_____

2. $4\dfrac{1}{2}$
$-1\dfrac{1}{3}$
_____

3. $8\dfrac{5}{9}$
$-3\dfrac{2}{9}$
_____

4. $16\dfrac{3}{10}$
$-5\dfrac{2}{5}$
_____

5. $1\dfrac{1}{2}$
$+3\dfrac{2}{3}$
_____

6. $4\dfrac{2}{3}$
$+3\dfrac{4}{5}$
_____

7. $6\dfrac{3}{4}$
$+9\dfrac{7}{8}$
_____

Solve.

8. $\dfrac{5}{16} \div \dfrac{3}{4} =$ _____

9. $\dfrac{1}{2} \times \dfrac{2}{3} =$ _____

10. $\dfrac{7}{8} \div \dfrac{5}{12} =$ _____

Change each mixed number to an improper fraction.

11. $9\frac{5}{8}$ = ——

12. $25\frac{2}{3}$ = ——

13. $10\frac{3}{4}$ = ——

Fill in the blanks. Write any remainders as fractions.

14. 14 qt = _____ gal

15. 8 gal = _____ qt

16. 20 pt = _____ qt

17. The area of a rectangle is 672 square yards. The width is 16 yards. What is the length?

18. Give the width of the rectangle in #17 in feet.

19. What is the GCF of 14 and 56?

20. Mental Math! 9 plus 1, minus 2, times 7, plus 4, equals _____ .

Divide using the reciprocal, then check your work by using the rule of four.

Reciprocal

Rule of Four

1. $\dfrac{3}{4} \div \dfrac{1}{2} =$

2. $\dfrac{3}{4} \div \dfrac{1}{2} =$

3. $\dfrac{4}{5} \div \dfrac{2}{3} =$

4. $\dfrac{4}{5} \div \dfrac{2}{3} =$

Change mixed numbers to improper fractions and divide using the reciprocal.

5. $1\dfrac{2}{3} \div \dfrac{5}{11} =$

6. $1\dfrac{7}{8} \div \dfrac{1}{8} =$

7. $2\dfrac{3}{4} \div \dfrac{1}{2} =$

8. $6\dfrac{1}{5} \div 3\dfrac{3}{8} =$

Solve.

9. $\begin{array}{r} 20\dfrac{1}{5} \\ - \ 10\dfrac{3}{4} \\ \hline \end{array}$

10. $\begin{array}{r} 9\dfrac{1}{8} \\ - \ 4\dfrac{1}{3} \\ \hline \end{array}$

11. $\begin{array}{r} 5\dfrac{3}{7} \\ + \ 5\dfrac{1}{4} \\ \hline \end{array}$

Fill in the blanks. Write any remainders as fractions.

12. 17 oz = ____ lb   13. 28 pt = ____ qt

14. 10 yd = ____ ft   15. 5 lb = ____ oz

16. 9 qt = ____ pt   17. 19 qt = ____ gal

18. Bria has 3 1/8 pounds of chocolates. If she divides them into portions that each weigh 5/8 of a pound, how many people can she treat?

19. Oliver's rectangular garden cart measures four feet by two feet by one foot. If he fills it to the top, how many cubic feet of mulch can he haul in one load?

20. Mental Math! 6 times 7, plus 3, divided by 9, plus 2, equals _____ .

Solve.

1. $3\frac{3}{8}$

    $+ 4\frac{1}{8}$

2. $7\frac{3}{5}$

    $+ 4\frac{4}{5}$

3. $15\frac{2}{7}$

    $+ 11\frac{1}{2}$

4. $8\frac{5}{6}$

    $+ 5\frac{2}{3}$

Solve.

5. $9\frac{7}{8}$

    $- 3\frac{5}{8}$

6. 3

    $- 1\frac{2}{5}$

7. $35\frac{1}{4}$

    $- 21\frac{2}{3}$

8. $7\frac{1}{6}$

    $- 6\frac{3}{7}$

Fill in the blanks. Write any remainders as fractions.

9. 48 oz = ____ lb

10. 23 pt = ____ qt

11. 9 yd = ____ ft

12. 4 lb = ____ oz

13. 16 qt = ____ pt

14. 20 qt = ____ gal

Divide using the reciprocal, then check your work by using the rule of four.

Reciprocal   Rule of Four

15. $\dfrac{3}{6} \div \dfrac{1}{2} =$

16. $\dfrac{3}{6} \div \dfrac{1}{2} =$

17. $\dfrac{5}{8} \div \dfrac{2}{3} =$

18. $\dfrac{5}{8} \div \dfrac{2}{3} =$

Change mixed numbers to improper fractions and divide using the reciprocal.

19. $6\dfrac{1}{3} \div 2\dfrac{3}{8} =$

20. $2\dfrac{3}{5} \div 1\dfrac{9}{10} =$

21. $5\dfrac{5}{6} \div \dfrac{5}{6} =$

22. $3\dfrac{1}{2} \div 1\dfrac{3}{7} =$

23. What is the reciprocal of 8?

24. A pool is 15 feet long, 10 feet wide, and 6 feet deep. How many cubic feet of water are required to fill it to the brim?

25. Joseph has 4 1/2 pizzas for his party. If there are eight people to serve, how much pizza can each one have?

26. Christopher jogged 2 1/3 miles and ran 1 5/6 miles. How far did he go altogether?

27. At the beginning of the day, Ethan had 5 1/2 rows to weed in his garden. If he has finished 2 2/3 rows, how much is left to do?

28. Mental Math! 8 times 6, minus 4, divided by 2, plus 3, equals _____ .

Use the multiplicative inverse to solve each equation. Check your work by replacing the unknown with the solution.

Solve.   Check.

1. 10J = 200   2. 10J = 200

3. 6U = 24   4. 6U = 24

5. 5K = 95   6. 5K = 95

Divide using the reciprocal, then check your work by using the rule of four.

Reciprocal   Rule of Four

7. $\dfrac{1}{3} \div \dfrac{2}{5} =$ 8. $\dfrac{1}{3} \div \dfrac{2}{5} =$

Change mixed numbers to improper fractions and divide using the reciprocal.

9. $1\dfrac{1}{4} \div 3\dfrac{1}{2} =$ 10. $4\dfrac{1}{6} \div 4\dfrac{4}{5} =$

Solve.

11. $3\dfrac{1}{3}$

       $-1\dfrac{2}{3}$

      _____

12. $8\dfrac{2}{3}$

       $+5\dfrac{3}{4}$

      _____

13. $4\dfrac{1}{6}$

       $-1\dfrac{1}{2}$

      _____

Solve.

14. $\dfrac{2}{3} \times \dfrac{3}{5} =$

15. $\dfrac{2}{7} \times \dfrac{3}{4} =$

16. $\dfrac{5}{8} \times \dfrac{1}{3} =$

17.   If Richard is six feet tall, how many inches tall is he?

18.   Three times Jacki's age is 120. How old is Jacki? Write an equation and solve.

19.   How many inches are there in one-half of a foot?

20.   A square measures two-thirds of a mile on each side. Give the area and the perimeter of the square.

Change any mixed numbers to improper fractions and multiply. Use canceling as a short cut.

1. $\dfrac{2}{3} \times \dfrac{3}{4} \times \dfrac{2}{7} =$

2. $1\dfrac{5}{7} \times \dfrac{1}{5} \times 1\dfrac{3}{4} =$

3. $\dfrac{1}{8} \times 1\dfrac{1}{7} \times \dfrac{4}{7} =$

4. $1\dfrac{1}{5} \times \dfrac{5}{6} \times 1\dfrac{1}{2} =$

Use the multiplicative inverse to solve each equation. Check your work.

Solve.   Check.

5. $7L = 77$ 6.

7. $13Y = 195$ 8.

Divide using whichever method you prefer.

9. $1\dfrac{9}{16} \div \dfrac{5}{8} =$

10. $6\dfrac{1}{3} \div 2\dfrac{3}{8} =$

Solve.

11. $6\frac{5}{8}$

    $- 3\frac{7}{8}$

12. $12\frac{1}{4}$

    $- 11\frac{3}{4}$

13. $1\frac{1}{3}$

    $+ 1\frac{2}{3}$

Fill in the blanks. Write any remainders as fractions.

14.   2 tons = _____ lb

15.   24 in = _____ ft

16.   1/4 ton = _____ lb

17.   Lillian's car weighs 1 1/2 tons. How many pounds does it weigh?

18. Five-eighths of the money was for food. One-tenth of that was used for eating out, and Rose got four-fifths of that amount for herself because she traveled a lot. What part of the total money did Rose get for eating out?

19. If 6 1/2 dozen donuts are divided among 13 people, how many donuts does each person receive?

20. Mental Math! 28 divided by 4, plus 1, times 8, minus 4, equals _____ .

Use the multiplicative inverse to solve each equation. Check your work.

Solve.   Check.

1. $6B + 10 = 40$ 2.

3. $8N - 13 = 3$ 4.

5. $6C + 8 = 50$ 6.

Change any mixed numbers to improper fractions and multiply. Use canceling as a short cut.

7. $1\dfrac{1}{5} \times 3\dfrac{1}{3} \times \dfrac{3}{11} =$ 8. $\dfrac{1}{2} \times \dfrac{5}{12} \times 3\dfrac{3}{7} =$

Divide using whichever method you prefer.

9. $\dfrac{1}{5} \div \dfrac{8}{25} =$ 10. $6\dfrac{2}{3} \div 2\dfrac{6}{7} =$

Fill in the blanks. Write any remainders as fractions.

11. 3 mi =_____ ft

12. 7 lb = _____ oz

13. 3 3/4 ft = _____ in

14. A farmer's truck was carrying 2 1/4 tons of hay and 3 1/4 tons of straw. How many tons were loaded on the truck?

15. Is 349 divisible by 9?

16. What are the prime factors of 48?

17. What is the GCF of 18 and 24?

18. A square measures 1 1/2 miles on each side. What is the area of the square?

19. Solve: 16^2

20. Mental Math! 56 divided by 7, plus 10, minus 3, plus 15, equals _____

Use the fractional value of π (22/7) to find the area and circumference of each circle.

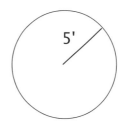

1. area = _____

3. area = _____

2. circumference = _____

4. circumference = _____

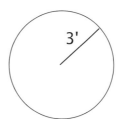

5. area = _____

7. area = _____

6. circumference = _____

8. circumference = _____

Use the multiplicative inverse to solve each equation. Check your work.

9. 5R – 17 = 38

10. Check

Solve.

11. $1\dfrac{1}{3} \times 1\dfrac{1}{2} =$

12. $1\dfrac{2}{3} \times 2\dfrac{1}{3} \times \dfrac{9}{10} =$

Solve.

13. $\dfrac{7}{10} \div \dfrac{7}{12} =$

14. $4\dfrac{2}{5} \div \dfrac{7}{9} =$

Use <, >, or = to compare the measures.

15. 19 qt $\bigcirc$ 5 gal

16. 3 tons $\bigcirc$ 5,000 lb

17. 2/3 mi $\bigcirc$ 3,720 ft

18. Keith drove 8 2/3 miles from home, then started to walk back. After he had walked 6 1/2 miles, how much farther did he have to travel to get home?

19. How many feet did Keith have left to travel (#18)?

20. The base of a triangle is 4 inches and the height is 2 1/2 inches. What is the area of the triangle?

28

Solve using the additive and multiplicative inverses as needed. Check your work.

1. $\dfrac{4}{15} W + 7 = 19$

2. Check

3. $\dfrac{2}{3} R - 5 = 5$

4. Check

5. $\dfrac{1}{2} X = 4$

6. Check

Change any mixed numbers to improper fractions and multiply.

7. $1\dfrac{7}{8} \times 2\dfrac{2}{3} =$

8. $\dfrac{4}{5} \times 3\dfrac{1}{3} \times 1\dfrac{7}{8} =$

Divide using whichever method you prefer.

9. $\dfrac{11}{18} \div \dfrac{3}{8} =$

10. $4\dfrac{1}{2} \div 2\dfrac{1}{2} =$

Use <, >, or = to compare the measures.

11. 3 lb $\bigcirc$ 32 oz

12. 1/5 mi $\bigcirc$ 1,056 ft

13. 1/4 gal $\bigcirc$ 2 qt

Find the average of each series of numbers.

14. 4, 2, 6, 8   15. 1, 3, 5, 7   16. 8, 10, 12

Average = _____   Average = _____   Average = _____

17. What is the area of a circle with a radius of 28 yards?

18. What is the circumference of the circle in #17?

19. What is the volume of a cube that measures 2 3/4 inches on each side?

20. Mental Math! 9 times 8, minus 2, divided by 10, times 4, equals _____ .

Change each fraction to an equivalent fraction in hundredths, then to a percent.

1. $\dfrac{5}{5} = \dfrac{}{100} = $ _____%

2. $\dfrac{3}{5} = \dfrac{}{100} = $ _____%

Write each fraction in hundredths. Then write it as a decimal and as a percent.

3. $\dfrac{4}{5} = \dfrac{}{100} = $ _____ $ = $ _____%

4. $\dfrac{3}{4} = \dfrac{}{100} = $ _____ $ = $ _____%

Solve using the additive and multiplicative inverses as needed. Check your work.

5. $\dfrac{3}{5}Y + 2 = 26$

6. Check

Solve.

7. $\begin{array}{r} 7\frac{2}{5} \\ -\ 2\frac{4}{5} \\ \hline \end{array}$

8. $\begin{array}{r} 9 \\ -\ 6\frac{3}{4} \\ \hline \end{array}$

9. $\begin{array}{r} 10\frac{9}{10} \\ +\ 4\frac{3}{8} \\ \hline \end{array}$

Write the Arabic numeral that corresponds to each Roman numeral.

10. X

11. XIV

12. XXII

13. XXXV

Write the Roman numeral that corresponds to each Arabic numeral.

14. 3

15. 17

16. 25

17. 19

18. Sandra earned money for chores. This week she earned the following amounts on different days: 15 dollars, 10 dollars, 8 dollars, and 25 dollars. What is the average amount she earned per day?

19. What are the area and circumference of a circle with a radius of 14 inches?

20. The base of a triangle is 16 feet and the height is 5 3/4 feet. What is the area of the triangle?

Use the multiplicative and additive inverses to solve each equation. Check your work.

1. $\dfrac{4}{9}W + \dfrac{1}{3} = \dfrac{2}{3}$

2. Check

3. $\dfrac{1}{2}K - \dfrac{3}{4} = 4\dfrac{1}{4}$

4. Check

Write each fraction in hundredths. Then write it as a decimal and as a percent.

5. $\dfrac{1}{5} = \dfrac{}{100} = $ _____ = _____%

6. $\dfrac{1}{4} = \dfrac{}{100} = $ _____ = _____%

Solve.

7. $1\dfrac{7}{8} \div \dfrac{5}{8} = $ _____

8. $1\dfrac{2}{8} \times 3\dfrac{1}{5} \times 5\dfrac{1}{5} = $ _____

9. $\begin{array}{r} 8\dfrac{2}{3} \\ -\ 4\dfrac{5}{6} \\ \hline \end{array}$

10. $\begin{array}{r} 11\dfrac{1}{2} \\ +\ 6\dfrac{5}{9} \\ \hline \end{array}$

Write the Arabic numeral that corresponds to each Roman numeral.

11. LI   12. CXV   13. MDC

Write the Roman numeral that corresponds to each Arabic numeral.

14. 11   15. 300   16. 1,700

17. How many feet are there in two miles?

18. How many pounds are there in three tons?

19. Mary's rectangular fish pond is 15 feet long, 10 feet wide and 3 1/2 feet deep. What is the volume of the water if the pond is filled to the top?

20. Solve for the unknown: 8X – 8 = 40

Solve for the unknown. Check your work.

1. $7L = 49$ 2. Check

3. $3R - 8 = 22$ 4. Check

5. $\dfrac{5}{8}X + 9 = 19$ 6. Check

7. $\dfrac{3}{10}Y - \dfrac{1}{5} = \dfrac{2}{5}$ 8. Check

9. $\dfrac{3}{4}T + \dfrac{3}{8} = 6\,\dfrac{3}{8}$ 10. Check

Solve.

11. $\dfrac{9}{10} \times \dfrac{5}{7} \times \dfrac{2}{3} =$ _____

12. $2\dfrac{1}{9} \times 2\dfrac{1}{4} \times 1\dfrac{1}{11} =$ _____

Write each fraction in hundredths. Then write it as a decimal and as a percent.

13. $\dfrac{3}{4} = \dfrac{}{100} =$ _____ $=$ _____%

14. $\dfrac{1}{2} = \dfrac{}{100} =$ _____ $=$ _____%

Use the fractional value of π (22/7) to find the area and circumference of the circle.

15. area = _____

16. circumference = _____

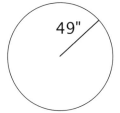

49"

Write the Arabic numeral that corresponds to each Roman numeral.

17. X   _____

18. I   _____

19. L   _____

20. M   _____

21. D   _____

22. V   _____

23. C   _____

Answer the questions.

24. Ron walked for five miles. How many feet did he walk?

25. Farmer Brown bought 6 1/2 tons of cattle feed. How many pounds of feed did he buy?

26. Eric is 5 3/4 feet tall. How many inches tall is he?

27. A room is 169 inches wide. How many feet wide is the room?

28. Wesley got the following scores on his unit tests: 100, 90, 85, 97. What is his average score?

Solve.

1. $\dfrac{1}{2}$ of 24 = ____

2. $\dfrac{2}{3}$ of 18 = ____

3. $\dfrac{7}{8}$ of 64 = ____

Fill in the missing numbers in the numerators or denominators to make equivalent fractions.

4. $\dfrac{3}{4} = \dfrac{}{} = \dfrac{}{} = \dfrac{}{16}$

5. $\dfrac{9}{10} = \dfrac{}{} = \dfrac{}{} = \dfrac{36}{}$

Use the rule of four to compare the fractions and write the correct symbol in the oval.

6. $\dfrac{5}{7}$ $\bigcirc$ $\dfrac{3}{5}$

7. $\dfrac{4}{8}$ $\bigcirc$ $\dfrac{3}{6}$

8. $\dfrac{4}{8}$ $\bigcirc$ $\dfrac{2}{3}$

Solve.

9. $\dfrac{3}{9} + \dfrac{5}{9} =$

10. $\dfrac{1}{2} + \dfrac{1}{4} + \dfrac{7}{8} =$

11. $\dfrac{4}{5} - \dfrac{1}{3} =$

12. $\dfrac{1}{3} \div \dfrac{1}{5} =$

13. $3\dfrac{1}{3} \div \dfrac{5}{18} =$

14. $3\dfrac{4}{5} \div 2\dfrac{7}{25} =$

Solve.

15. $\begin{array}{r} 7\frac{1}{4} \\ -\,5\frac{3}{4} \\ \hline \end{array}$

16. $\begin{array}{r} 9\frac{2}{3} \\ +\,6\frac{5}{9} \\ \hline \end{array}$

17. $\begin{array}{r} 5\frac{1}{5} \\ -\,2\frac{5}{6} \\ \hline \end{array}$

Write the length of the line.

18. _____ "

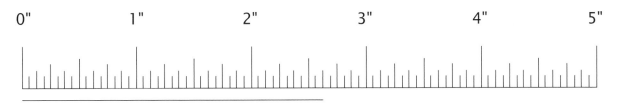

Solve for the unknown and check your work.

19. $7X + 9 = 44$

20. Check

21. $\dfrac{3}{8} A - 8 = 13$

22. Check

23. $\dfrac{5}{6} G + \dfrac{1}{6} = \dfrac{5}{12}$

24. Check

Solve.

25. $\dfrac{5}{8} \times \dfrac{1}{3} \times \dfrac{3}{5} =$ _____

26. $\dfrac{4}{5} \times 2\dfrac{3}{4} \times 3\dfrac{1}{3} =$ _____

Write each fraction in hundredths. Then write it as a decimal and as a percent.

27. $\dfrac{4}{5} = \dfrac{}{100} =$ _____ $=$ _____%

28. $\dfrac{1}{4} = \dfrac{}{100} =$ _____ $=$ _____%

29. What is the GCF of 15 and 45?

30. What are the prime factors of 56?

31. Change to an improper fraction: 7 2/3

32. Is 498 divisible by 9?

33. What is the area of a circle with a radius of 21 feet?

34. What is the circumference of a circle with a radius of 21 feet?